THE GREAT BOOK OF OPTICAL ILLUSIONS

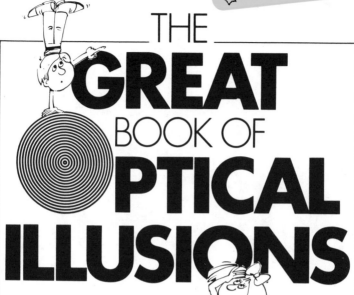

Gyles Brandreth

 Sterling Publishing Co., Inc.   New York

Sterling ISBN 0-8069-1767-9

Published in 1985 by Sterling Publishing Co., Inc.
387 Park Avenue South, New York, N.Y. 10016
Distributed in Canada by Oak Tree Press Ltd.
% Canadian Manda Group, P.O. Box 920, Station U
Toronto, Ontario, Canada M8Z 5P9
First American edition published in hardcover under the title
"Seeing Is Not Believing" © 1980 by Sterling Publishing Co., Inc.
Original edition published in the U.K. in 1979 under the title
"The Big Book of Optical Illusions" by Carousel Books, a division
of Transworld Publishers Ltd., text copyright © 1979
by Gyles Brandreth, illustrations copyright © 1979
by Transworld Publishers Ltd.

STOP!

Life is full of surprises—and so is this book. In it you're going to find all sorts of things that aren't quite what they seem.

Is Seeing Believing?

To begin with, take a close look at this shape:

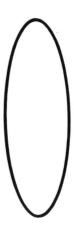

Do you agree that no one could possibly describe that shape as a perfect circle? Good. Now look at the picture of the bicycle on the next page. How would you describe the shape of the bicycle's front wheel? As a circle? Of course, because you *know* that a bicycle wheel has a circular shape.

The artist drew exactly the same shape above—that obviously isn't a circle—as he did to make the bicycle wheel—that obviously *is* a circle! But while you are looking at the same shape on both this page and the next, you realize that the bicycle wheel has to be circular even when it doesn't *seem* to be.

What I'm talking about is a very complicated phenomenon that is called **perception**. Perception is turning what we see into what we understand. Our perception is our view of the world. It's our perception that tells us the first shape looks a little like an elongated egg and the second looks like the circular wheel of a bicycle.

It's our perception that tells us that the word SUNSHINE is written here—

—when actually it isn't.

You were able to read the word SUNSHINE even though all the letters weren't written there—and they really weren't—because the black shadows made the word appear. The shadows enabled you to **perceive** the word, in the same way that these few shadows enable you to perceive the clear shape of a bird:

Usually, we see what we expect to see, but sometimes our perception lets us down and we perceive something to be so that isn't actually so!

That should happen to you quite a few times as you look through this book, because optical illusions very often manage to fool your perception and make you begin to *wonder if seeing is believing after all!*

Now You See It . . .

What's this?

An elegant vase or two old men?

It's both, of course. Concentrate on the white area and you'll see the vase. Concentrate on the black and the two men will appear.

Great Gaze

Gaze at the pattern on the facing page
for at least a minute.

Quickly look at a perfectly blank wall
and you'll find that you start seeing
strange little moving specks on the wall.

It's weird, but don't let it frighten you!

Five Fields

Here are five fields.

Which is the largest and which is the smallest in area?

All five fields have exactly the same area. The different shapes make them appear to be different sizes.

Blockbuster

This white block is a little bigger

than this white block

—isn't it?

14

Eye Teaser

Which is bigger: A or B?

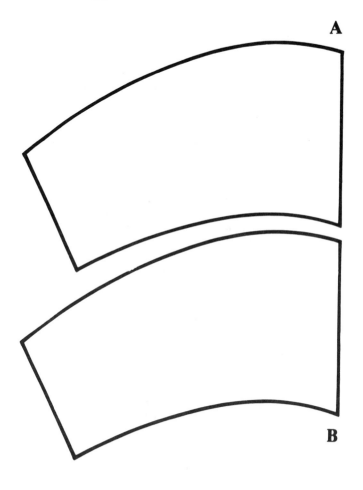

Wrong again! They're both the same size.

Curious Cube

Look carefully at the drawing on the next page and then try to answer these three questions:

1. Is the cube on a table and you are looking at it from above?
2. Is the cube in mid-air and you are looking at it from below?
3. Does the line across the corner of the cube seem slightly bent?

*You could be looking at the cube from above **or** below! Sometimes you will feel you're looking up at the cube from below and sometimes you will feel you are looking down on the cube from above.*

The line does seem to be slightly bent, but it isn't. It's perfectly straight!

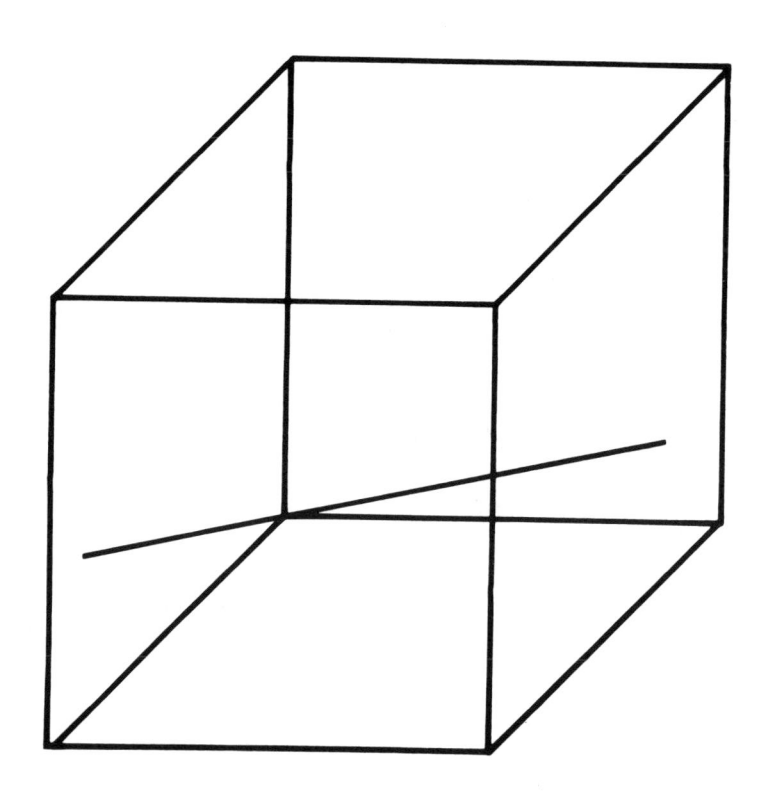

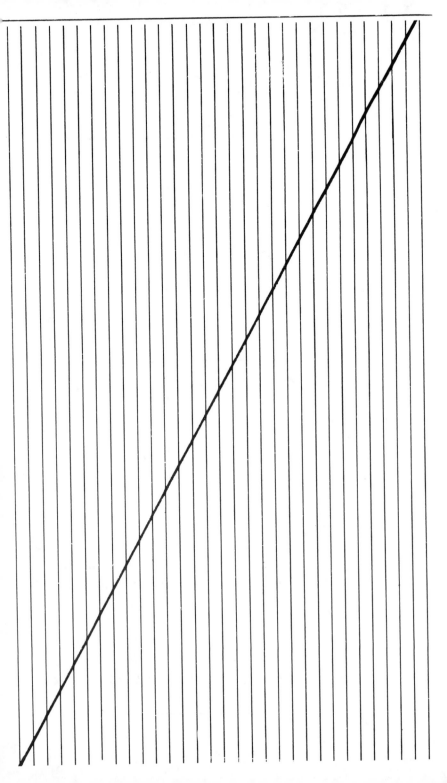

Ins and Outs

Look carefully at the diagonal line on the facing page.

Is it straight? Or does it twist in and out of the horizontal lines and seem a little jagged?

The diagonal line is quite straight. The vertical lines behind it make it seem distorted.

Around and Around

Look at *either* the circle on this page or the next.

Concentrate on it and revolve the book. Turn it around and around as quickly as you can.

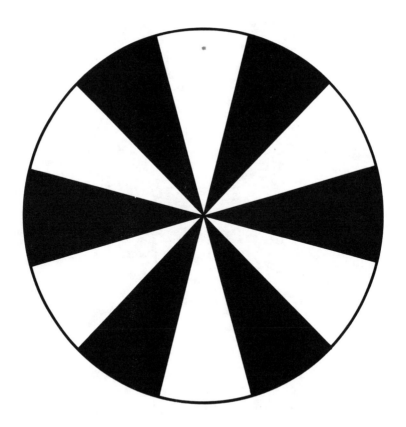

When you stop, for a moment the pattern will suddenly seem to go around in the opposite direction.

X Marks the Spot

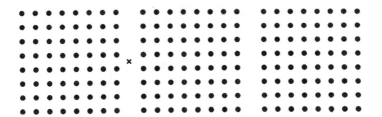

Focus on the spot marked X and you will find that the dots in the square on the left appear in horizontal **rows** while the dots in the square on the right appear in vertical **columns**.

It will always happen that way—never the other way around!

Right Angle?

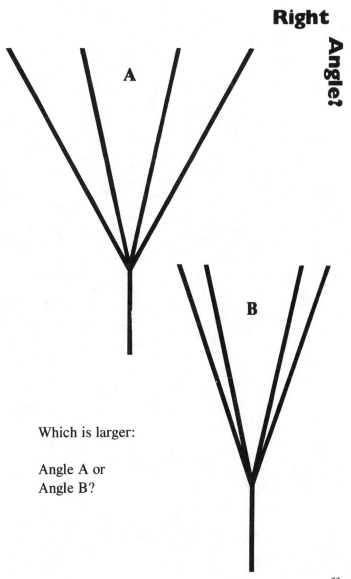

A

B

Which is larger:

Angle A or
Angle B?

The angles are the same! They look unequal because of the other angles on either side of them which are different.

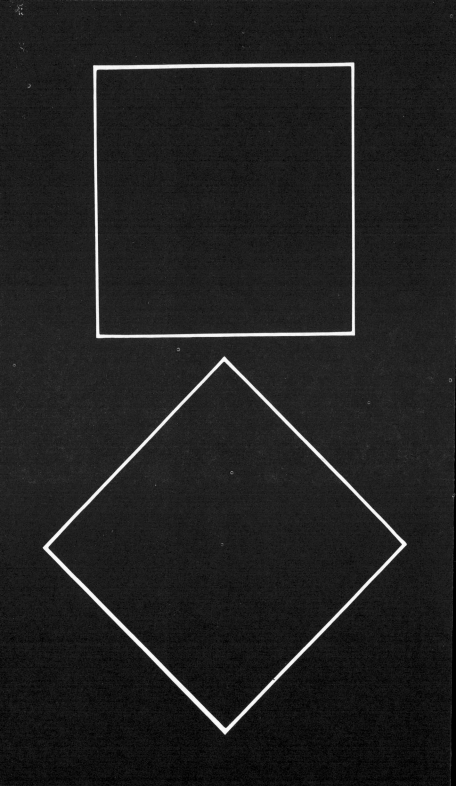

Diamonds and Squares

Which is bigger:

the diamond or the square?

They are both squares and are the same size, but a square will seem larger when tilted on one side and viewed like a diamond.

Eye Dazzler

Look at the facing page for long enough and your mind will really begin to boggle.

What can you see? Rows of triangles? Rows of squares? Rows of open boxes seen from above? Or a mixture of different patterns that keep changing as you look at them?

Far Out

Is B closer to A or closer to C?

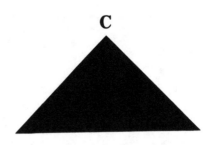

B looks a lot closer to A, doesn't it? In fact, B is exactly midway between A and C so the distance between A and B and B and C is identical.

Whazzit?

What's this a picture of?

*There's more to the picture than two lines, two black spots and a curve. The picture shows a cleaning lady on her knees scrubbing the floor with her bucket of water beside her. Now you **know** what's in the picture you'll have no problem seeing it, will you?*

Square World

Is one of these two areas very slightly
larger than the other?

Which one?

The white square seems a little larger than the black square, but in fact they're both exactly the same size.

Watch 'Em Bend!

Look at the star
on this page
steadily while
you count to 100
very slowly.

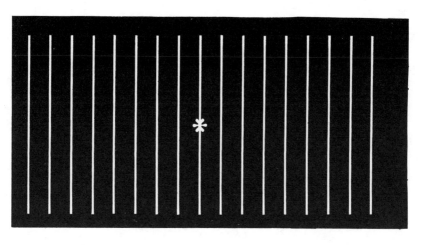

Now look at the
star on *this* page
and watch the
lines curve in
the opposite
direction!

33

Impossible!

Look at the opposite page.

Whichever way you look at it this is an "impossible" object.

That is, it is possible to *draw* it on paper, but you could never *build* it out of cardboard or wood.

If it looks perfectly all right to you, look again—starting at the base of the object and then letting your eye move up it.

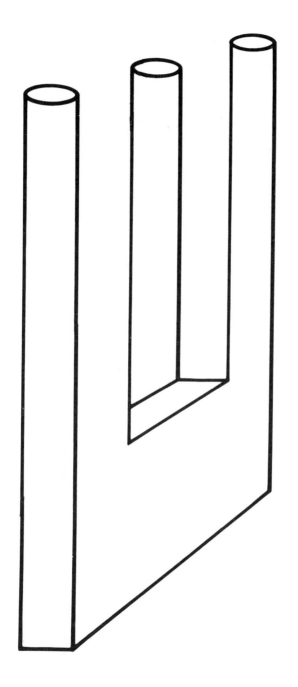

A Question of Lines

Which of the three horizontal lines is the longest: the top one, the middle one, or the bottom one?

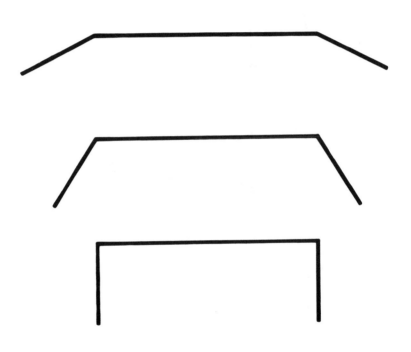

All three are the same length. It is the angles that make the horizontal lines look different lengths.

A Question of Angles

Which is the longer line:
the one from A to C
or
the one from B to D?

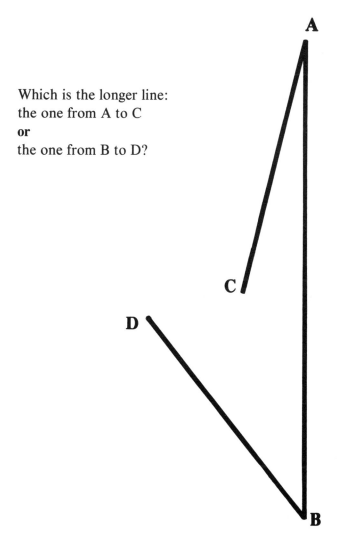

*Both lines are the same length. It is the **angles** the lines make that cause them to look different.*

Up or Down?

You will have to turn the book sideways to see this strange construction properly. And when you do look at it, are you seeing it from above or from below?

This is one of those odd figures that you feel you are seeing from above at one moment and from below at the next. Whichever way you look at it, it's still confusing!

39

Portrait of a Lady

What can you see
in this picture?

Is it a drawing of a very old lady?

Or is it a picture of a pretty girl with her
head turned slightly away from you?

It's both. Look at the picture long enough and you'll see
the old lady at one moment and the girl the next.

Bull's-Eye

Revolve the pages and the spirals will seem to get bigger or smaller depending upon in which direction you are turning the book.

Center Point

Glance at the facing page

and tell what you can see

right in the very middle of it.

If you said you saw the letter B that's because you first perceived the horizontal line of letters A, B, C. If you said you saw the number 13 that's because you first perceived the vertical line of numbers 12, 13, 14.

12
A13C
14

Seeing Circles

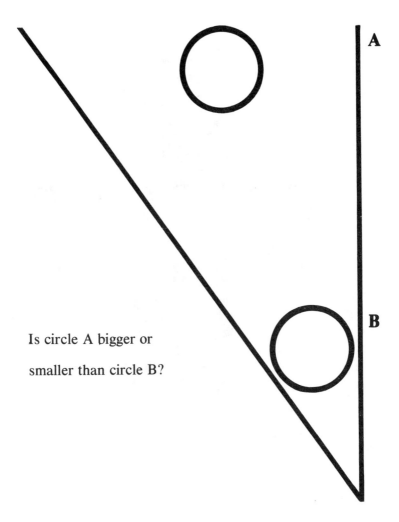

Is circle A bigger or

smaller than circle B?

Both the circles are the same size. If they look different,
it is because of their positions inside the angle.

46

Upstairs Downstairs

Find the top step. When (and **if**)
you find it, start looking for the
bottom step!

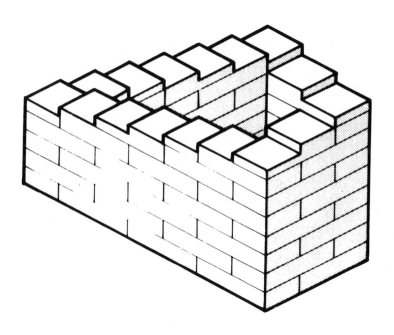

Straight and Narrow

How many of the vertical lines on the facing page are bending this way and that? And how many of them are perfectly straight?

They are all perfectly straight. It is the pattern of wavy lines behind them that makes them appear to bend.

How Far This Time?

Is the distance between A and B greater or smaller than the distance between C and D?

A _____

_____

_____

_____

_____

_____

C _____

_____

_____

_____

_____

_____

_____

_____

B

It looks greater, but in fact it's the same.

D _____

Master Carpenter

Ask a friend if he can build this hollow crate for you from 12 pieces of wood. Tell him he can have $1,000 if he succeeds!

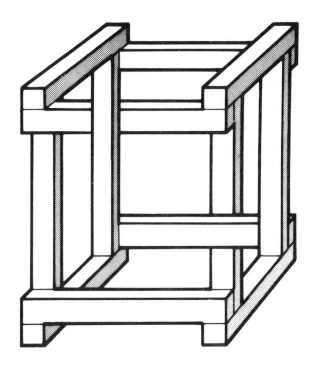

Some Circles

Of the two center circles, which one is the bigger?

They are both the same size. The top one only looks bigger because it is surrounded by smaller circles and the bottom one only looks smaller because it is surrounded by bigger circles.

Point of View

Look carefully at the
two horizontal lines.

Which one is longer: the
top one or the bottom
one?

*Both are exactly the same length. It's the difference in
position that makes the lower one look longer.*

A Curve Ball

Which of the three arcs is the biggest:

> the top one,
>
> the middle one,
>
> or the bottom one?

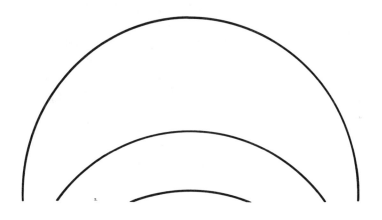

55

Countdown

How many cubes can
you count here?

The answer is 6 or 7. It will be 6 if you saw the patterned part as the top of each cube, but 7 if you saw the patterned part as the bottom of each cube.

And if you found that mind-boggling, you can really get yourself confused by trying to find your way through the cube maze. Using tracing paper, go in at one arrow and come out at the other.

(See page 95 for solution.)

In or Out?

Look carefully at the two vertical lines on the opposite page.

Do they bend inwards at the middle?

Do they bend outwards at the middle?

Are they perfectly straight?

*The lines **look** as if they bend inwards at the middle, but they don't. They are perfectly straight parallel lines.*

From Here to There

Is the line from A to B longer or shorter than the line from C to D?

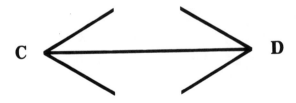

It looks longer, but both lines are actually the same length.

Where Are You?

Are you up in the sky looking down on the roof of a house? Or are you in a room looking into a corner?

*The answer is "either"! The line in the middle will either appear nearer to you (the roof) or farther away from you (the corner)—depending on how **you** see it!*

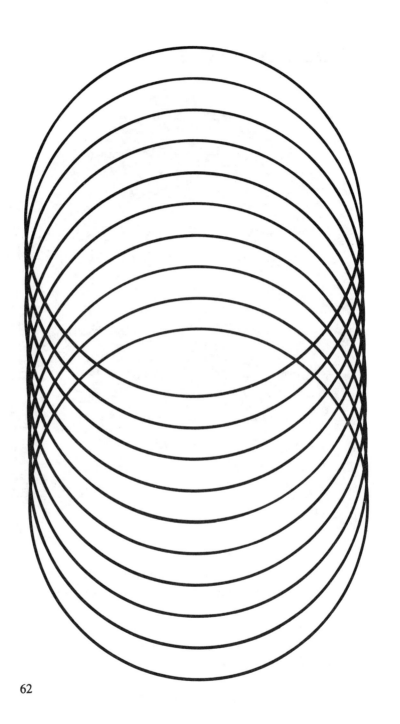

Which End Is Up?

Here's an unusual tube. Look carefully at it for at least a minute and then decide if you are looking down the tube from above it **or** up the tube from under it.

You can look at the tube either way. Sometimes you'll feel you're seeing it through it from the top and sometimes from the bottom!

Great or Small?

Which of the two circles is the larger?

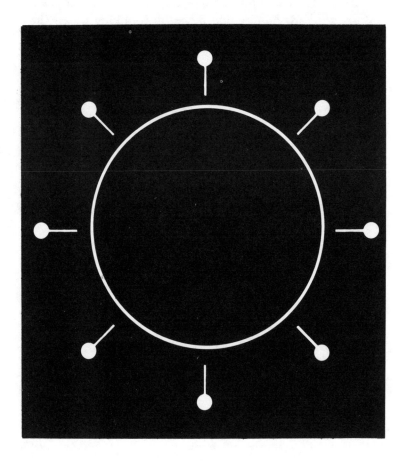

The one on this page?

Or the one on this page?

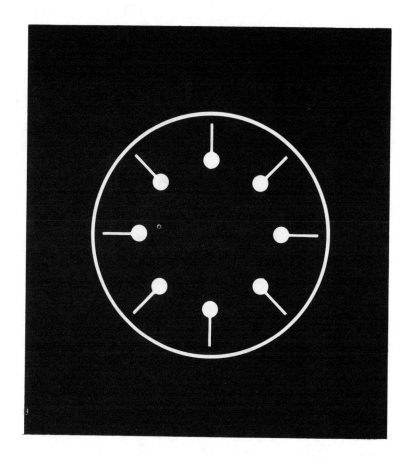

*You may be wrong! The one on this page certainly **looks** smaller, but in fact both circles are identical in size.*

Crisscross

Glance at the facing page and strange gray spots will appear at all the points where the lines cross. Look at any one crossing in particular and the gray spot that was there will suddenly dissappear!

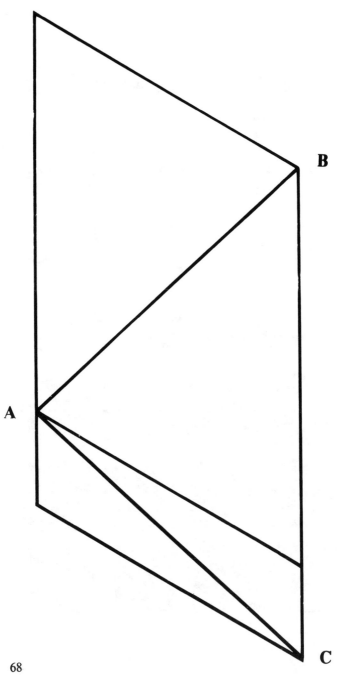

Parallel Bars

Look carefully at the parallelogram opposite.

Which line is longer: AB or AC?

AB looks a lot longer, but in fact both lines are the same length. Check with a ruler if you don't believe us.

Strange Circles

Which of the three rings is a perfect circle?

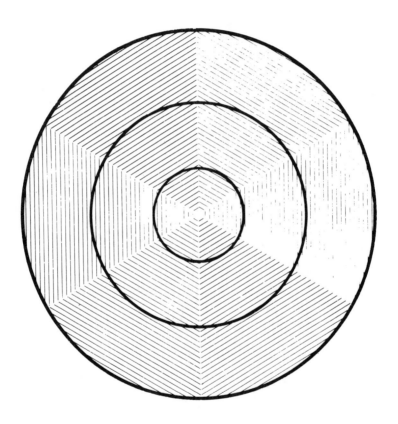

All three are! It is the background pattern that makes the perfect circles seem distorted.

Topsy

Turvy

Look carefully at this picture by the great Dutch artist Verbeek—then turn it upside down and give yourself a surprise.

The largest of the Rocs picks her up by the skirt.

Just as he reaches a small grassy point of land, another fish attacks him, lashing furiously with his tail.

Fair and Square

Of the three squares,
which one is the
smallest?

All three squares are identical. The ones with the vertical and horizontal lines in them just seem to occupy a
larger area.

Revolver

Look at the circle on the opposite page and keep looking at it.

As you look at it, it will seem to revolve.
(Don't look at it for too long, or you might begin to feel a little dizzy!)

Thick and Thin

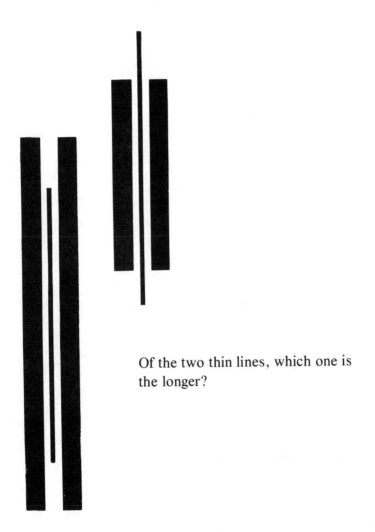

Of the two thin lines, which one is the longer?

The two thin lines are of equal lengths. They look differ-ent because of the different lengths of thick lines on either side of them.

Square Dance

Why does each side of the square bend inwards slightly in the middle?

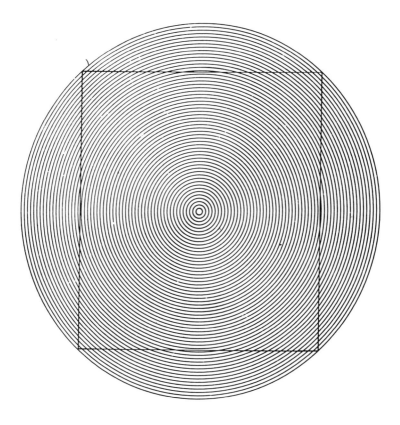

*In fact, it's a perfect square. The sides don't bend at all. They just **seem to** because of the pattern of rings behind.*

Curvy Lines

Look carefully at the two vertical lines.

Do they bend outwards in the middle?

Do they bend inwards in the middle?

They don't bend at all! They are perfectly straight parallel lines.

Wooden Triangle

If you're any good at carpentry
try making this simple wooden triangle.

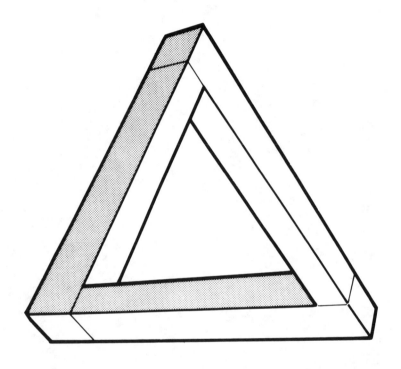

Actually, you can't do it! The wooden triangle is one of those "impossible objects" that are easy to draw, but not to make.

Honeycomb

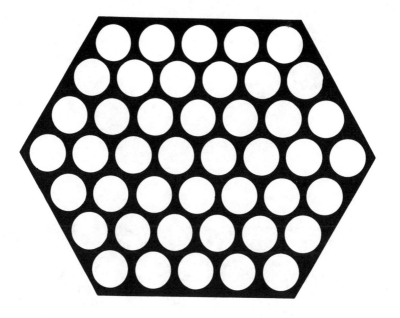

Look at this pattern long enough

and you'll find the circles begin

to look like hexagons!

82

Gray Matter

Look at the three shades of gray in the circle
and two rectangles and decide which of the
three is darkest.

*The bottom rectangle looks darkest because it is set
against a white background, but in fact all three gray
areas are identical.*

Square Circles

Which of the two circles is the larger?

A

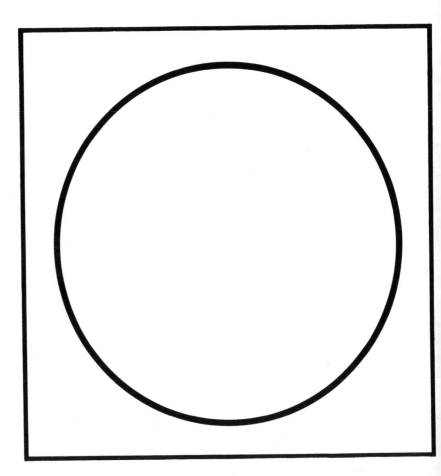

A **looks** a little larger than B, but in fact both the circles are the same size. It's the position and size of the boxes that deceive you.

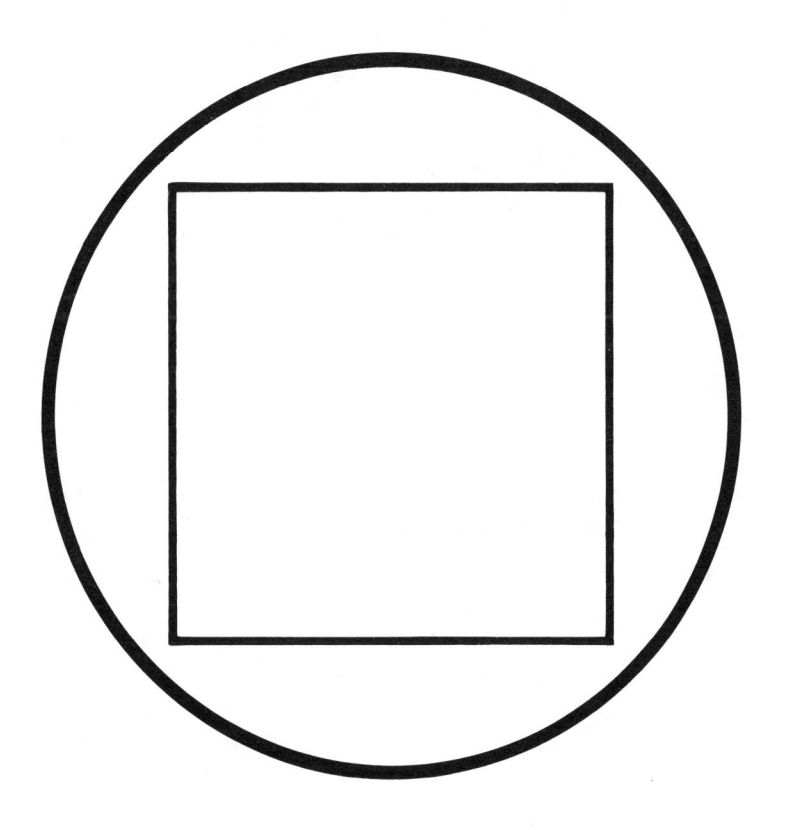

B

Staircase

Look at these stairs any way you like and you'll still feel you could climb them. Whether you look at the page as it is or turn it sideways or turn it upside down, you'll still find stairs to climb.

It's amazing—so amazing, in fact, that it's a maze as well as an eye-teaser. Using tracing paper, go in at one arrow and see how long it takes you to come out at the other.

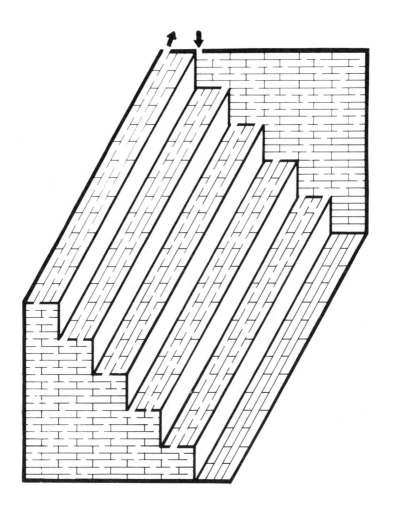

(See page 95 for solution.)

The Long and Short of It

Which of the two horizontal lines is the longer? It looks like the top one, but are you **sure**?

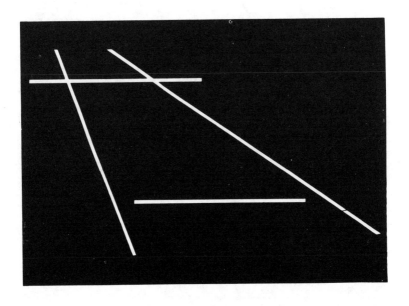

Both the horizontal lines are the same length. It's the converging lines that make the top line look longer.

Animal Magic

What creature have we here?

If you said a rabbit you were right.

If you said a duck you were right too!

It all depends on whether the left part of the picture strikes you as being a duck's bill before it strikes you as being a rabbit's ears.

Amazing Hat

Here's a high hat.

How much greater is its height than
its width?

The height of the hat and the width of the brim are identical—vertical lines often seem longer than horizontal lines of the same length. And it's not just an amazing optical illusion, it's an amazing maze as well. Using tracing paper, try going in at one arrow and coming out at the other. *(See page 95 for solution.)*

Tile and Error

Look at this wall of black and white tiles.

Why do you think the rows aren't straight?

They are! The vertical rows of tiles are perfectly straight. Admittedly they don't look it, but put a ruler to them and you'll see they are.

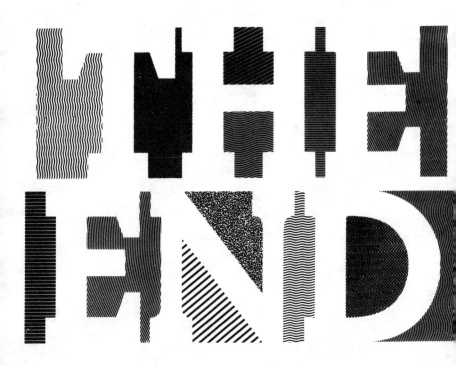

Maze Solutions

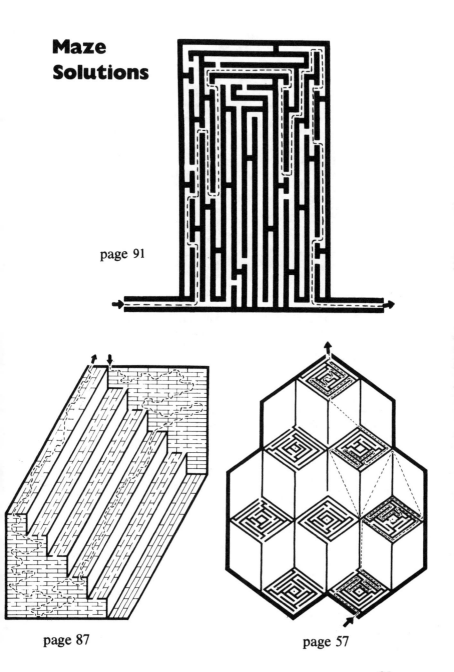

page 91

page 87

page 57

Index